THE SCIENCE DIPLOMACY PLAYBOOK

Daniella Sussman

THE SCIENCE DIPLOMACY PLAYBOOK

2026 EDITION

Published by The Global Lens: Science Diplomacy in Focus

WWW.GLSD.AI · Washington, DC

For permissions and licensing inquiries: daniella@globallenspodcast.com

ABOUT THE AUTHOR

Daniella Sussman works at the intersection of science, innovation, and global diplomacy, advising leaders as emerging technologies reshape geopolitical and research ecosystems. She supports governments, industry, and research institutions on strategic positioning, identifying areas of leverage and building trusted international partnerships.

She translates complex scientific and technological developments into clear, actionable insights for policy and executive decision-making, strengthening alignment across academia, industry, and government while advancing global research engagement and technology leadership.

She brings more than two decades of experience across the United Nations system, in senior advisory roles, with work spanning Africa, Central Asia, the Middle East, and Europe, as well as global engagement across 193 countries.

She is the founder of The Global Lens: Science Diplomacy in Focus and the GLSD.AI platform, editor of Global Signals, host of The Global Lens podcast, and a board member of the Global Network for Science and Diplomacy.

FOREWORD: WHY THIS BOOK EXISTS

Science diplomacy has a communications problem. The practitioners who understand it best rarely write for general audiences. The scholars who write about it tend toward abstraction. And the government officials who exercise it daily operate under constraints that prevent frank disclosure.

The result is a growing field with a huge gap between enormous stakes and insufficient practical guidance.

This playbook exists to close that gap. It is written for the practitioner: the science attaché navigating a bilateral negotiation, the research institution building its first international partnership, the foreign ministry official suddenly responsible for technology policy, the investor trying to decode geopolitical risk before a capital commitment.

Science diplomacy is no longer a niche concern. It is the operating environment for major power competition, industrial strategy, technology governance, and international research. The leaders who understand it will be positioned to shape the next decade. Those who do not will find themselves responding to decisions made by others.

CONTENTS

INTRODUCTION

> *The transition happened faster than most institutions noticed. What had been a niche domain for research partnership agreements became the primary arena for great power competition.*

In 2015, if you described science diplomacy as a strategic priority to a foreign ministry official, you might have been redirected to the cultural attaché. By 2023, the same conversation was happening at the cabinet level. By 2025, it was embedded in industrial strategy documents, export control frameworks, and national security directives.

The shift was driven by several converging forces. First, the US-China technology competition forced a reckoning that had been deferred for decades: technological leadership is not separable from economic and military power. Semiconductors, AI infrastructure, and telecommunications became the terrain of strategic competition before most foreign ministries had updated their frameworks to reflect it. Second, the COVID-19 pandemic exposed the vulnerability of nations that had offshored scientific and manufacturing capability. Third, the clean energy transition created a new resource map. Lithium, cobalt, nickel, and rare earth elements are not distributed evenly. The nations that control them, process them, or access them through partnership hold structural advantage.

The practitioners who understand this new landscape operate across multiple registers simultaneously: scientific, diplomatic, commercial, and strategic. This playbook is designed to give you fluency in all four.

CHAPTER 01:

THE ARCHITECTURE OF SCIENCE DIPLOMACY

> *Before you can navigate the field, you need to understand its structure. Science diplomacy is not a single domain. It is a family of practices, each with different rules, different actors, and different leverage points.*

THREE MODES OF SCIENCE DIPLOMACY

The Royal Society and the American Association for the Advancement of Science (AAAS) introduced a framework in 2010 that remains useful: science in diplomacy, diplomacy for science, and science for diplomacy. Understanding which mode you are operating in determines your tools.

Science In Diplomacy

Science in diplomacy integrates scientific knowledge into foreign policy: science advisors embedded in foreign ministries, technical experts in negotiating delegations, and scientific evidence shaping treaty positions. The Paris Agreement is the canonical example.

Diplomacy For Science

Diplomacy for science uses diplomatic tools and government channels to facilitate international scientific cooperation. The leverage point is access: data sharing, facility access, researcher mobility, and stable funding flows. Each can be enabled or obstructed by diplomatic relationships.

Science For Diplomacy

Science for diplomacy uses scientific cooperation as a diplomatic tool. Nobel Laureate Dr. Peter Agre's work offers compelling evidence that the most important conversations in global diplomacy are not always happening in foreign ministries. Sometimes they happen in a laboratory in Pyongyang, a research institute in Havana, or a conference room in Tehran, between scientists who share a language that moves across political divides.

THE THREE INSTITUTIONAL LAYERS

Intergovernmental

The UN system, multilateral development banks, treaty bodies. Sets norms, establishes frameworks, allocates resources at global scale. Slow by design, broad in scope, often weak in enforcement.

Bilateral

Science and technology agreements between countries, joint research programs, and data sharing arrangements. Examples include the US-India Science and Technology Endowment Fund, which supports joint applied research and commercialization, or the Australia-Japan quantum computing partnership. Where most substantive work happens; these are faster, more specific, and directly consequential for commercial and strategic outcomes.

Institutional

Universities, research councils, national laboratories operating with significant autonomy. These are science diplomacy actors in their own right, and where some of the most consequential failures have occurred, as institutions have built international programs without understanding the geopolitical risks they were accumulating.

> The practitioner who can move fluently across all three layers is rare and valuable. Forums like the Science Diplomacy Summit at Johns Hopkins SAIS exist precisely to develop this kind of practitioner, who elevates emerging voices, forges multi-stakeholder networks, and presents new models of collaboration.

PUT IT TO WORK

- Identify which of the three modes best describes your current work, then assess whether you are under-leveraging the other two.
- Map which institutional layer you primarily operate in. Name one person or institution you should know in each of the other two layers that you do not currently have a relationship with and write down how you will make contact.
- List the three most consequential science and technology developments affecting your domain right now. For each, identify which institutional layer and mode of science diplomacy is most relevant.
- Review your organization's international S&T activities and categorize each by mode and layer. Identify where you have concentration risk.

CHAPTER 02:

MAPPING THE PLAYERS

> *In science diplomacy, the formal hierarchy rarely matches the actual power map. Understanding where decisions are made, and by whom, is the first operational skill.*

FOREIGN MINISTRIES AND SCIENCE ATTACHÉS

Most foreign ministries have scientific and technological affairs units, but their influence varies enormously. Before engaging, understand the internal architecture. Who reports to whom? Does the science advisor have access to the political leadership? Is the technical agenda driving policy, or is it subordinated to commercial or security priorities?

RESEARCH FUNDING AGENCIES

National research funding agencies shape the international science landscape through grant-making, partnership requirements, and research priority-setting. They are no longer merely scientific: the recent trend toward conditionality in research funding has made these agencies explicitly political.

NATIONAL LABORATORIES

National laboratories operate at the intersection of scientific research, national security, and industrial policy. They are often the most technically sophisticated actors in any science diplomacy negotiation and the most constrained by security classifications and export controls. The best access points are through open science programs and academic partnerships.

THE PRIVATE SECTOR

Technology companies, pharmaceutical firms, energy majors, and defense contractors are increasingly central. The semiconductor industry is clarifying: Taiwan Semiconductor Manufacturing Company (TSMC) is a Taiwanese company with global strategic importance. The US CHIPS Act, the EU Chips Act, and equivalent initiatives in Japan, South Korea, and India are all attempts to reshape the geography of semiconductor manufacturing, which is fundamentally a science diplomacy challenge dressed in industrial policy language.

CIVIL SOCIETY AND SCIENTIFIC COMMUNITIES

Scientific communities exercise significant soft power by setting norms for open science, establishing standards for research integrity, and creating channels for cooperation that operate below the political radar. The Intergovernmental Panel on Climate Change (IPCC) has shaped climate negotiations for three decades on this basis.

> Power in science diplomacy is not always where the org chart says it is. The most consequential actors often operate through informal networks and technical committees that have no direct reporting line to a political principal.

PUT IT TO WORK

- For your top two or three priority countries, map the actual science and technology decision-makers, not just the formal org chart. Identify who controls the budget and who the minister actually listens to.
- Identify the national research funding agency most relevant to your domain in each key partner country. Find the name of the program officer who covers your area.
- List the three most influential scientific communities in your field. Assess your organization's standing with each. Where you have no standing, identify the fastest credible path in.
- Identify one private sector actor whose technology or capital decisions are shaping your domain. Map their government relationships and assess whether your organization has a point of engagement.

CHAPTER 03:

THE BILATERAL PLAYBOOK

> *Bilateral science and technology agreements are the workhorses of science diplomacy. Building, maintaining, and leveraging them requires specific skills.*

ANATOMY OF A SCIENCE AND TECHNOLOGY AGREEMENT

A bilateral S&T agreement typically contains a statement of purpose and scope, mechanisms for joint funding and cost-sharing, intellectual property provisions, data sharing and publication rights, personnel exchange and visa facilitation, dispute resolution mechanisms, and provisions for review and renewal. Recent examples include the US-South Korea Technology Prosperity Deal, which focused on semiconductors, AI, and quantum, and various Germany-India agreements focused on critical minerals, semiconductors, and green hydrogen.

The most important elements are the ones most negotiators spend the least time on: IP provisions and data sharing. In an era of intense competition over emerging technology, an agreement that does not clearly specify who owns what, and what restrictions apply to that ownership, is a liability.

THE NEGOTIATION PROCESS

Bilateral Science and Technology (S&T) negotiations typically begin with a memorandum of intent, proceed through working-level technical discussions, and culminate in a formal agreement signed at an appropriate political level. The political level of the signing matters; it signals the strategic importance both parties attach to the relationship.

MANAGING THE PARTNERSHIP THROUGH ITS LIFECYCLE

Agreements get signed. Programs get launched. And then, in many cases, the partnership drifts. Active management, tracking deliverables, maintaining political relationships on both sides, and flagging emerging tensions before they become crises, are disciplines most institutions underinvest in.

WHEN PARTNERSHIPS BREAK DOWN

Bilateral scientific partnerships are vulnerable to political disruptions that neither party controls. The practitioner's job in these situations is triage: identify what can be preserved, maintain communication channels even when formal cooperation has been suspended, and position the partnership for rapid restoration when the political environment improves.

> The best bilateral partnerships have champions on both sides who are invested in the relationship's success and have the standing to defend it when it comes under political pressure. Building those champions is a strategic priority, not a nice-to-have.

PUT IT TO WORK

- Audit your organization's three most important bilateral S&T partnerships. For each: Are IP provisions clearly specified? Is there an active management structure? Are there named champions on both sides?
- Identify your single most important bilateral partnership. Map the political and institutional relationships that sustain it. If one of those relationships disappeared tomorrow, what would happen?
- For any partnership showing signs of drift, develop a re-engagement plan with one specific deliverable, a timeline, and a named relationship to reactivate.
- Draft a one-page partnership health brief for your most important bilateral relationship. Update it every six months.

CHAPTER 04:

MULTILATERAL MECHANISMS IN A FRACTURED WORLD

> *The UN system remains the foundational architecture of international science cooperation. But the architecture is under stress, and new structures are being built alongside it. The practitioner who understands only the UN system is navigating with an incomplete map.*

THE UN SCIENCE SYSTEM: STILL ESSENTIAL, BUT UNDER PRESSURE

The UN system's science and technology infrastructure spans multiple agencies: the United Nations Educational, Scientific and Cultural Organization (UNESCO) for education, science, and culture; the World Health Organization (WHO) for health science and biosecurity; the International Atomic Energy Agency (IAEA) for nuclear technology; the World Meteorological Organization (WMO) for climate and atmospheric science; the Food and Agriculture Organization (FAO) for agricultural science; and the United Nations Environment Programme (UNEP) for environmental science. Cutting across these is the UN's new Global Dialogue on AI Governance, formally established by General Assembly resolution in August 2025 and launched at a high-level multi-stakeholder meeting on 25 September 2025. It is designed to connect national, regional, and sectoral AI initiatives and promote coherence across governance regimes. Its first session is scheduled for 6 to 7 July 2026 in Geneva.

However, 2026 is also a year of institutional strain. The UN faces a leadership transition, with the selection process for the next

Secretary-General (to take office in January 2027) underway under conditions of fiscal constraint and political fragmentation. A reform effort, the UN80 Initiative, launched by the Secretary-General in March 2025, reflects an attempt to modernize how the organization operates, focusing on efficiencies, mandate reviews, and operational improvements rather than any grand redesign of multilateralism. The US has strongly opposed centralized multilateral AI governance initiatives, declining to sign the Paris AI Summit declaration in February 2025 and expressing firm rejection of global AI oversight efforts, including at the UN Security Council. This signals a structural shift in how the world's leading AI power engages with collective governance.

The implication for practitioners is not that the UN system is irrelevant. It is that UN-anchored outcomes increasingly depend on who is in the room and who is not, and that the absence of major powers from key processes shapes outcomes as much as their presence.

HOW DECISIONS ACTUALLY GET MADE

In theory, UN decision-making emphasizes consensus and transparency. In practice, outcomes are shaped before they reach the floor of any intergovernmental body. The regional group system, including the Africa Group, Group of 77 (G77) and China, European Union, and the Western European and Others Group (WEOG), is the primary vehicle for position coordination. If you want to influence the outcome of a multilateral negotiation, you need to engage with the relevant regional groups before the formal session begins.

UN secretariats are more powerful than their formal mandates suggest. They control the agenda, produce the documentation, draft the negotiating texts, and manage the information flows that shape outcomes. Building relationships with secretariat staff, particularly the technical officers who do the substantive work, remains one of the highest-leverage activities available to a science diplomacy practitioner.

THE RISE OF MINILATERALISM

The most consequential shift in the multilateral landscape is not happening inside the UN. It is happening in the proliferation of small, issue-specific coalitions of like-minded states, what scholars and practitioners now call minilateralism.

The logic is straightforward: traditional multilateral institutions operate by consensus across large memberships, which creates chronic gridlock in policy outcomes; the tyranny of the lowest common denominator. Minilateral groupings, by contrast, bring together a small number of states with aligned interests to move faster, go deeper, and make binding commitments that larger forums cannot reach.

The Quadrilateral Security Dialogue (Quad), comprising the United States, Japan, India, and Australia, is the most prominent example. Revived in 2017 and elevated to leaders-level summits, it has expanded well beyond its initial maritime security focus to cover critical minerals, clean energy, semiconductor supply chains, next-generation telecommunications, and health security. Its Quad Critical and Emerging Technology Working Group is, in effect, conducting science diplomacy at the highest political level, outside any UN framework.

AUKUS, the trilateral security pact among Australia, the United Kingdom, and the United States, represents the hard-power complement to the Quad's normative agenda. Pillar II of AUKUS covers artificial intelligence, quantum computing, cyber capabilities, hypersonic weapons, and undersea systems. Consultations and invitations for Japan, Canada, and New

Zealand to participate in select Pillar II projects on a project-by-project basis are ongoing, gradually expanding this minilateral into a broader technology alliance.

Other significant minilateral structures include the US-Japan-India Trilateral Strategic Dialogue; the Indo-Pacific Economic Framework (IPEF); and the Minerals Security Partnership. Each operates on the same logic: a coalition of willing, aligned states moving faster on specific issues than universal multilateral bodies can manage.

AI GOVERNANCE: THE FRAGMENTATION TEST CASE

Nowhere is the tension between multilateralism and minilateralism more acute than in artificial intelligence governance. The UN's Global Dialogue on AI Governance, established by the General Assembly in August 2025, is designed to provide an inclusive, stable home for AI governance coordination. Its stated goals are to build safe, secure, and trustworthy AI systems grounded in international law; to promote interoperability between governance regimes; and to encourage open innovation accessible to all.

The challenge is that the three largest AI powers are pursuing fundamentally different governance models. The United States, under the current administration, has explicitly rejected centralized control and global governance of AI, withdrawing from or opposing multilateral AI governance compacts and positioning itself as a promoter of American standards rather than a participant in collective frameworks. The European Union has enacted the AI Act, a comprehensive risk-based regulatory framework that is already shaping global markets through its extraterritorial reach. China has expressed strong support for global governance frameworks, particularly those that give developing countries a seat at the table. At the same time, it is actively building bilateral AI partnerships across the Global South.

The practical result is a fragmented governance landscape: three competing AI stacks, three different regulatory philosophies, and a UN-anchored dialogue that the largest AI power has declined to support. For science diplomacy practitioners, this fragmentation is not a problem to solve. It is the operating terrain which they must navigate.

WHAT THE NEW ARCHITECTURE MEANS FOR PRACTITIONERS

The shift from universal multilateralism to a hybrid system of multilateral frameworks, minilateral coalitions, and bilateral partnerships has specific implications for how science diplomacy is practiced.

Bilateral partnerships remain foundational. Examples include the US-India Science and Technology Endowment Fund for joint applied research and commercialization, and the Germany-India cooperation on semiconductors, critical minerals, hydrogen, and renewable energy.

Minilateral formats such as the Quad's Critical and Emerging Technology Working Group or AUKUS Pillar II on advanced capabilities including AI and quantum have become increasingly important in science diplomacy, technology, and security. These smaller groupings allow faster progress on emerging issues like AI governance, quantum computing, semiconductors, and critical minerals supply chains, where universal forums often face gridlock.

Trusted partner status has become a prerequisite for access to the most consequential technology cooperation. Export controls, technology transfer restrictions, and alliance-based industrial policy are all mechanisms for defining who is inside and who is outside the trusted partner perimeter. Countries and institutions that have not invested in the relationships, agreements, and interoperability standards that signal trustworthiness will find themselves excluded from the science and technology partnerships that matter most.

Minilateral forums require different engagement strategies than multilateral ones. There are no regional group coalitions to build, no secretariat relationships to cultivate, no rules of procedure to master. Access depends on direct relationships with member state officials and on demonstrated value to the specific agenda the minilateral is pursuing. The practitioner who approaches a Quad working group with UN negotiating tactics will fail. The practitioner who understands what each member state is trying to achieve and what they can contribute will be effective.

The Global South faces a structural challenge in this new architecture. Minilateral coalitions are exclusive by design. Countries outside the trusted partner perimeter have less access to leading-edge technology cooperation and less influence over the standards and norms being set. China's strategy of building AI partnerships with Global Majority countries through the UN framework, positioning itself as the advocate for inclusive governance against an exclusive Western coalition, is partly a response to this dynamic.

FINANCING SCIENCE: THE UN MECHANISMS STILL MATTER

Despite the structural shifts in governance, UN financing mechanisms remain essential for many science diplomacy actors, particularly in the Global South. The Global Environment Facility, the Green Climate Fund, the Technology Mechanism of the United Nations Framework Convention on Climate Change (UNFCCC), and the Climate Technology Centre and Network (CTCN) all channel significant resources to science and technology programs. The UN's Independent International Scientific Panel on AI, established by the same August 2025 General Assembly resolution (with nominations and work advancing in 2026), represents a new channel for science-policy linkage at the global level.

Project proposals that arrive without institutional backing rarely succeed, regardless of technical merit. Building the institutional relationships that support successful proposals remains as important as it has ever been.

> The multilateral system is not being replaced. It is being supplemented, competed with, and in some domains bypassed. The practitioner who navigates only the UN system is working with an outdated playbook. The practitioner who understands the full architecture, universal multilaterals, regional organizations, minilateral coalitions, bilateral partnerships, and private sector platforms, and who knows when to use which, will be the one who gets things done.

PUT IT TO WORK

- Map the full governance architecture in your domain: which issues are governed by UN frameworks, which by minilateral coalitions, which by bilateral agreements, and which by private sector standards bodies. Identify where the gaps are and where the action is happening.
- Assess your organization or country's status in the trusted partner frameworks most relevant to your domain. Are you inside the perimeter on critical minerals, AI cooperation, semiconductor access, or space governance? If not, what is the path in?
- Identify one minilateral forum relevant to your work. Research its membership, agenda, and governance structure. Identify the specific official in each member state who covers your issue. These are your access points.
- For any UN financing mechanism you are pursuing, map the political relationships that support successful applications. Technical merit is necessary but not sufficient. Identify who needs to advocate for your proposal before it reaches the formal review process.
- Track the AI governance fragmentation actively. Know where your key partner countries stand relative to the US, EU, and Chinese regulatory models. The gap between governance frameworks creates both risks and opportunities for organizations that can operate across them.

CHAPTER 05:

CRITICAL TECHNOLOGIES AND THE NEW DIPLOMACY

> *Technology has always been a subject of diplomacy. What is new is the speed, the breadth, and the stakes. The technologies at the center of current competition will determine the balance of power for decades.*

ARTIFICIAL INTELLIGENCE: THE GOVERNANCE FRONTIER

AI governance has become one of the most contested domains in international affairs. AI is no longer simply a technology domain. It is emerging as a geopolitical actor, reshaping statecraft, enabling new forms of economic competition, and raising questions that no existing multilateral framework is equipped to answer.

Key forums include the Organisation for Economic Co-operation and Development (OECD) AI Policy Observatory, the Global Partnership on AI, the UN AI Advisory Body, the International Telecommunication Union (ITU)'s AI for Good platform, and bilateral dialogues including the US-EU Trade and Technology Council and the US-Japan tech partnership.

BIOTECHNOLOGY AND BIOSECURITY

The pandemic demonstrated that biosecurity is a global public good requiring international scientific cooperation to produce and sustain. The Biological Weapons Convention lacks a verification mechanism. The emerging agenda, covering pandemic prevention architecture, dual-use research oversight, and synthetic biology governance, is technically complex with enormous strategic stakes.

QUANTUM TECHNOLOGIES

Quantum computing, communications, and sensing represent a genuinely transformative frontier with direct national security implications. UK-US quantum collaborations are among those shaping foundational frameworks for next-generation technologies. The science diplomacy opportunity is primarily at the standards level.

SPACE AND SCIENCE DIPLOMACY

Space is simultaneously a domain of international scientific cooperation and major power competition. The Artemis Accords have created a parallel governance structure to the UN Committee on the Peaceful Uses of Outer Space. Commercial space companies are conducting activities with significant diplomatic implications and limited formal oversight.

> Technology governance is science diplomacy's growth market. The frameworks being built now will govern technologies that will be consequential for fifty years. The practitioners who engage today will be disproportionately positioned to shape outcomes.

PUT IT TO WORK

- Select the critical technology domain most consequential for your work. Identify the three governance forums where the rules are currently being negotiated and assess your organization's standing in each.
- Assess your organization's AI governance engagement. If you are not represented in the relevant forums, identify who from your organization would best fill this role, what credentials they need, and what timeline is realistic.
- For each technology domain you work in, map the export control and regulatory landscape in your key partner countries. Identify where restrictions create operational risk.
- Identify one technology standards body relevant to your domain and find out whether your organization has a path to formal participation or observer status.

CHAPTER 06:

CRITICAL MINERALS, ENERGY, AND RESOURCE DIPLOMACY

> *The clean energy transition has created a new resource map. Understanding it is not optional for anyone operating in science diplomacy, international development, or strategic affairs.*

THE CRITICAL MINERALS LANDSCAPE

Rare earth elements can be divided into two categories with different strategic profiles. Light rare earths are more abundant and found in many parts of the world. Heavy rare earths, critical for high-performance magnets in defense systems, electric vehicle motors, and precision guidance, are far less common and far more strategically sensitive.

The concentration of power lies not primarily in where materials are found but in processing, separation, metal-making, and magnet production: capabilities that require specialized expertise, infrastructure, and scale that remain highly concentrated.

SUPPLY CHAINS AS STRATEGIC SYSTEMS

For decades, mineral supply chains were designed for efficiency under an assumption of stability. That assumption no longer holds. U.S. export controls are interpreted by China within a much wider strategic frame. China has responded by using its position in critical minerals and processing as counter-leverage, with export restrictions on gallium and germanium being the clearest example. A fully healthy market is not defined by lowest cost or maximum efficiency. It is defined by its ability to balance competitiveness with resilience.

THE DIPLOMATIC RESPONSE

The Minerals Security Partnership coordinates investment in mineral-rich developing countries and shares supply chain risk information. What is emerging is not simply diversification but system design: distributed but aligned ecosystems across trusted partners. USA Rare Earth's approach illustrates this, with capabilities in the United Kingdom through Less Common Metals and collaboration in France through Carester to expand heavy rare earth processing capacity.

Bilateral partnerships remain essential building blocks within this broader diplomatic response. Recent examples include the US-DRC Strategic Partnership Agreement, signed in December 2025, which focuses on investment, local processing capacity, supply chain security, and preferential access to critical mineral assets, as well as Japan's substantial support for the Nacala Corridor infrastructure and minerals development across Mozambique, Malawi, and Zambia, which emphasizes technology transfer, value addition, and integrated logistics on African terms.

One of the least visible but most critical challenges is workforce development. The capabilities required in rare earth processing depend on specialized, experience-based knowledge that has eroded outside China over decades. Capital alone does not solve this.

The same transition driving critical mineral demand is reshaping energy infrastructure diplomacy. Advanced reactor designs, green hydrogen, and next-generation energy systems all require international scientific cooperation at scale. Australia, Chile,

Morocco, and Namibia are positioning as major green hydrogen exporters, and the trade and cooperation frameworks to govern these flows are being built now alongside the mineral partnerships that will supply them.

> Critical minerals are not a commodity story. They are a power story. The institution that understands the full supply chain, from extraction through processing, manufacturing, and recycling, and has relationships across that chain, will be positioned to navigate the transition.

PUT IT TO WORK

- Map the critical minerals relevant to your sector or jurisdiction. For each, trace the supply chain: primary production, processing, manufacturing, and end use. Mark every concentration point.
- Identify your single most vulnerable supply chain dependency. Assess whether diversification is feasible in your timeframe and what diplomatic or commercial relationship would reduce the exposure.
- Review the Minerals Security Partnership member list and the EU Critical Raw Materials Act. Identify where your organization's interests align with existing frameworks and where you have gaps.
- For any resource diplomacy or energy infrastructure initiative your organization is considering, assess whether cooperation frameworks are in place, and whether commitments to processing capacity, workforce development, and technology transfer are included. If not, reconsider whether it is durable.

CHAPTER 07:

THE AFRICAN RESOURCE SOVEREIGNTY SHIFT

> *The framing has changed. Africa's position in the global science and technology order is no longer defined by what external partners are willing to offer. It is defined by what African governments are willing to accept.*

GEOLOGICAL LEVERAGE IS REAL AND BEING USED

Africa holds approximately 30 percent of the world's critical mineral reserves, including cobalt, lithium, manganese, graphite, platinum group metals, and bauxite. More specifically, the DRC produces more than 70 percent of global cobalt, Zimbabwe holds some of the world's largest lithium deposits, South Africa dominates platinum group metal production and Namibia is emerging as a significant uranium mining destination. Beyond minerals, Namibia and Morocco are also positioning as major green hydrogen exporters, adding an energy dimension to Africa's strategic resource leverage.

This endowment has translated directly into diplomatic leverage, and African governments are using it. The DRC has imposed restrictions on raw cobalt exports to push downstream processing onto African soil, moving from an outright export ban to a strict quota system that controls global supply volumes through 2027. Zimbabwe has done the same with lithium. The African Union's Agenda 2063 explicitly names mineral value chains as a continental strategic priority, not a development aspiration.

BENEFICIATION DEMAND IS NON-NEGOTIABLE

Beneficiation, the requirement that minerals be processed domestically before export, is now a central plank of mineral policy across the African continent. Countries that export raw ore while importing processed materials are, in effect, exporting leverage. The countries and institutions that are serious about African partnerships are arriving with technical training programs, joint processing investments, and co-development commitments.

Effective bilateral partnerships in this environment are increasingly expected to include concrete commitments to local processing, workforce development, and technology transfer on African terms. Recent examples include the US-DRC Strategic Partnership Agreement of December 2025, which links investment to value addition and local industrialization; Japan's Nacala Corridor initiatives across Mozambique, Malawi, and Zambia, with significant infrastructure and technical cooperation funding; and growing South Africa-DRC cooperation on mining, energy, and logistics value chains.

SCIENTIFIC CAPACITY ON AFRICAN TERMS

African scientific capacity is growing, unevenly but with clear direction. South Africa, Egypt, Kenya, Nigeria, Ethiopia, and Morocco have research universities, national science councils, and S&T policy infrastructure capable of genuine peer collaboration. The relevant practitioner question is not what African institutions need but what they are building, and whether you can contribute on their terms.

> Africa's resource sovereignty is not a diplomatic complication to be managed; it is the new configuration of power in this field. The practitioner who grasps that African governments are negotiating from a position of geological leverage, and engages accordingly, will be effective.

PUT IT TO WORK

- Before any new engagement in Africa, conduct a country-specific assessment covering: mineral endowment and current beneficiation policy, scientific capacity and institutional landscape, and the political context for international partnerships.
- Audit your organization's current Africa engagement model. For each active partnership, answer honestly: Does it include processing capacity development? Workforce training? African-led research design?
- Identify three African research institutions, universities, or policy bodies most relevant to your domain. Assess which you have active relationships with and which represent untapped potential.
- Map the beneficiation policies in force or under development in the countries most relevant to your work. If your engagement does not account for these policies, you are operating on an outdated map.

CHAPTER 08:

INSTITUTIONAL STRATEGY

> *Science diplomacy is practiced by individuals operating through institutions. Building the institutional platform that gives you the standing to operate effectively requires deliberate strategy.*

UNDERSTANDING INSTITUTIONAL LEGITIMACY

In science diplomacy, legitimacy comes from multiple sources: scientific credentials, institutional affiliation, government mandate, and demonstrated track record. Most practitioners have some combination of these and need to understand which they have and how to deploy them.

BUILDING YOUR NETWORK

Science diplomacy operates through networks. The practitioner who knows the right people in the foreign ministry, the research council, the relevant multilateral secretariat, and the private sector will consistently outperform the practitioner who has better formal qualifications but weaker relationships.

Building this network requires consistent presence in the relevant spaces: the right conferences, the right working groups, the right informal gatherings. It requires genuine investment in relationships: following up, providing value, showing up when it matters. Trust is cultivated over time, through consistent interactions, and cannot be accelerated.

INTELLIGENCE AND POSITIONING

Effective science diplomacy practitioners maintain active intelligence about their environment: who is doing what, who has resources, where opportunities are emerging, what the political dynamics are in the countries and institutions they work with. The practitioner who is consistently well-informed has a durable competitive advantage.

> Institutional strategy is not separate from science diplomacy work. It is the prerequisite for it. Without a credible platform, you cannot convene, you cannot broker, you cannot influence. Build your platform deliberately.

PUT IT TO WORK

- Conduct a personal legitimacy audit. List your scientific credentials, institutional affiliations, government mandates, and track record of delivered commitments. Identify your strongest asset and the most significant gap.
- Map your current professional network against the key nodes in your domain: foreign ministry contacts, research council relationships, multilateral secretariat connections, and private sector links. Identify the three most important gaps.
- Develop a 90-day relationship-building plan. Include at least one new relationship in each of the three institutional layers, with a specific forum, event, or introduction as the entry point.
- Identify the one person in your network whose endorsement would most increase your credibility with a key audience you cannot currently reach. Invest in that relationship before you need to call on it.

CHAPTER 09:

THE INTELLIGENCE ADVANTAGE

> *The practitioner who understands the landscape before others do is positioned to act before others can. Strategic intelligence is not a luxury; it is a core professional competency.*

WHAT STRATEGIC INTELLIGENCE MEANS

Strategic intelligence in the science diplomacy context means knowing what governments, research institutions, and technology companies are doing and planning; understanding how geopolitical dynamics are shaping scientific cooperation; tracking the emerging technology landscape; and anticipating how regulatory and policy changes will affect partnerships and programs.

BUILDING YOUR INTELLIGENCE SYSTEM

Effective intelligence systems typically combine systematic monitoring of official sources, active engagement with professional networks, regular engagement with scientific literature in relevant domains, and investment in analytical tools that help manage information volume.

The Global Signals intelligence brief, published biweekly by GLSD.ai, is designed to support practitioners who need to track this landscape without the resources to build a full intelligence operation. It synthesizes the most consequential developments across science diplomacy, emerging technology, and international affairs.

USING INTELLIGENCE EFFECTIVELY

The most effective use involves identifying opportunities before they are widely recognized, anticipating risks before they materialize, and building the conversations and relationships that position you to act when opportunities emerge.

> Intelligence is only valuable when it changes your decisions. Build your intelligence system around the specific domains where you need to act, and maintain the discipline to act on what you learn.

PUT IT TO WORK

- Audit your current information sources against your actual intelligence requirements. For each priority in your work, ask: do I have a reliable source for early signals in this domain?
- Establish a weekly intelligence review practice. Set aside 30 minutes to scan your curated sources with one specific question in mind: What has changed this week that affects a current priority?
- Subscribe to Global Signals and identify two additional primary sources, one official government or multilateral document stream and one expert network, to add to your monitoring stack.
- For the most important opportunity or risk you are currently tracking, write a one-paragraph intelligence summary and share it with one colleague. The discipline of writing it will sharpen your analysis.

CHAPTER 10:

THE PRACTITIONER'S TOOLKIT

> *Science diplomacy requires a specific set of skills, each of which can be developed with deliberate practice.*

TRANSLATION AND COMMUNICATION

The ability to communicate clearly across scientific, diplomatic, and policy communities is the foundational skill of science diplomacy. The practitioner who can write a briefing note that accurately represents scientific findings, in language that a minister can act on, in a format that the diplomatic community will find credible, has a skill that is genuinely rare and consistently undervalued.

NEGOTIATION

Science diplomacy involves negotiation at every level: over research priorities, over IP provisions in agreements, over the language in international declarations, over the allocation of resources in joint programs. The specific techniques of multilateral negotiation, including constructing coalitions, managing procedural leverage, bridging positions, and drafting compromise text, can be learned.

CULTURAL AND POLITICAL LITERACY

A negotiating position that makes perfect sense in a Washington framework may be politically impossible for a counterpart operating within African Union, Chinese, or EU constraints. Building this literacy requires time in country, language study, and genuine investment in understanding how your partners see the world.

STRATEGIC COMMUNICATION

The practitioner who can explain why a particular bilateral S&T agreement matters, what it delivers in concrete terms, and how it serves the interests of both parties has a significant advantage in securing resources and political support.

> The most effective science diplomacy practitioners are not necessarily the best scientists or the most experienced diplomats. They are the people who operate fluently across scientific, diplomatic, and political registers, and who never stop building the relationships that make things happen.

PUT IT TO WORK

- Assess your current capability across the four skills: translation and communication, negotiation, cultural and political literacy, and strategic communication. Identify your weakest area and name one concrete action you can take to develop it.
- Take one key message from your current work and draft it three ways: for a scientific audience, for a diplomatic audience, and for a political audience.
- Identify one upcoming negotiation or working group session. Before it, write down the other party's interests, constraints, and red lines as you understand them. After the meeting, assess how accurate you were.
- Find one senior science diplomat whose career you want to study and request a conversation. Most will say yes. Bring specific questions. The relationship will pay dividends for years.

CONCLUSION: THE FIELD IS MOVING. YOU SHOULD BE TOO.

Science diplomacy is not a static field, and this playbook is not a finished map. The terrain is shifting fast. The emergence of AI as a diplomatic and industrial priority was barely visible five years ago. The critical minerals competition has intensified in the last three years. The space for scientific cooperation between major powers has contracted while the importance of that cooperation has grown.

The practitioners who will be effective in this environment share certain characteristics: they invest in understanding rather than just information, they build relationships rather than just contacts, they think about institutional strategy deliberately, and they maintain genuine intellectual curiosity about the scientific and technological developments that are reshaping global power.

Science diplomacy is ultimately about people. The agreements, the frameworks and the multilateral mechanisms, matter, but they are created and sustained by human relationships, by trust built through repeated interactions, and by commitments made and kept.

The stakes are high; the world needs more people who can do this work well.

REFERENCES

African Union Commission. 2013. Agenda 2063: The Africa We Want. Addis Ababa: African Union Commission.

Agre, Peter. 2025. Can Scientists Succeed Where Politicians Fail? Baltimore: Johns Hopkins University Press.

Bisley, Nick. 2025. "The Quad, AUKUS and Australian Security Minilateralism." The Pacific Review. https://doi.org/10.1080/10670564.2024.2365241

Duarte, Rui Pedro. 2025. Statecraft 3.0: The Age of AI Diplomacy. Self-published.

European Commission. 2024. Regulation (EU) 2024/1689 Laying Down Harmonised Rules on Artificial Intelligence (Artificial Intelligence Act). Official Journal of the European Union.

International Energy Agency (IEA). 2023. Critical Minerals Market Review. Paris: IEA.

Japan International Cooperation Agency (JICA). 2025-2026. Nacala Corridor Infrastructure and Minerals Development Initiative: Mozambique, Malawi, and Zambia. Progress reports.

Kissinger, Henry, Eric Schmidt, and Daniel Huttenlocher. 2021. The Age of AI: And Our Human Future. London: John Murray.

Lee, Kai-Fu. 2018. AI Superpowers: China, Silicon Valley, and the New World Order. Boston: Houghton Mifflin Harcourt.

Miller, Chris. 2022. Chip War: The Fight for the World's Most Critical Technology. New York: Scribner.

Naim, Moises. 2009. "Minilateralism: The Magic Number to Get Real Results." Foreign Policy, June 21.

NASA. 2020. The Artemis Accords: Principles for Cooperation in the Civil Exploration and Use of the Moon, Mars, Comets, and Asteroids. Washington, DC: NASA.

Organisation for Economic Co-operation and Development (OECD). 2019. OECD Principles on Artificial Intelligence. Paris: OECD.

Royal Society and American Association for the Advancement of Science (AAAS). 2010. New Frontiers in Science Diplomacy. London and Washington, DC: Royal Society and AAAS.

Royal Society and American Association for the Advancement of Science (AAAS). 2025. Science Diplomacy in an Era of Disruption. London and Washington, DC: Royal Society and AAAS.

United Nations Educational, Scientific and Cultural Organization (UNESCO). 2021. Recommendation on the Ethics of Artificial Intelligence. Paris: UNESCO.

United Nations Environment Programme (UNEP) and African Union. 2025. Critical Minerals for Africa's Development: Value Addition and Beneficiation Strategies.

United Nations Framework Convention on Climate Change (UNFCCC). 2023. Technology Mechanism and Climate Technology Centre and Network (CTCN) Reports.

United Nations General Assembly. 2025. Resolution A/RES/79/325: Terms of Reference and Modalities for the Establishment and Functioning of the Independent International Scientific Panel on Artificial Intelligence and the Global Dialogue on Artificial Intelligence Governance. 27 August 2025.

U.S. Department of State. 2025. U.S.-DRC Strategic Partnership Agreement on Critical Minerals. December 2025.

U.S. Department of State. 2026. Minerals Security Partnership (MSP) Ministerial Outcomes. February 2026.

White House. 2025. Executive Order 14179: Removing Barriers to American Leadership in Artificial Intelligence. 23 January 2025.

World Bank. 2023. Minerals for Climate Action: The Mineral Intensity of the Clean Energy Transition. Washington, DC: World Bank.

FURTHER READING

Brookings Institution. 2026. Foresight Africa 2026. Washington, DC: Brookings Institution.

International Monetary Fund (IMF). 2024. World Economic Outlook: Navigating Global Transitions. Washington, DC: IMF.

World Economic Forum (WEF). 2025. Global Risks Report 2025. Geneva: WEF.

Additional official documents, including UN resolutions, EU AI Act implementation materials, OECD updates, and Minerals Security Partnership statements, are referenced throughout the text or available through their respective institutional platforms.

GLOBAL SIGNALS THE GLOBAL LENS PODCAST SIGNALS LIVE BRIEFINGS

GLSD.AI

DANIELLA SUSSMAN

STRATEGIC ADVISORY · SCIENCE DIPLOMACY · GLOBAL PARTNERSHIPS

WASHINGTON, DC | AVAILABLE WORLDWIDE | GLSD.AI

For institutions navigating science diplomacy and emerging technology at a strategic level.

WORK WITH DANIELLA

Three ways to engage, depending on where you are and what you need.

GOVERNMENT ADVISORY

Strategic counsel for foreign ministries, intergovernmental bodies, and science and technology agencies on technology governance, research partnerships, and industrial strategy positioning.

EXECUTIVE BRIEFINGS

Targeted intelligence sessions for leadership teams on science diplomacy developments, emerging technology trends, geopolitical risk, and the policy and partnership implications that matter to your sector.

INSTITUTIONAL STRATEGY

Advisory support for research institutions, universities, multilateral organizations, and industry leaders developing international partnerships and navigating complex multi-stakeholder environments.

SHORT-TERM PROJECT-BASED · ONGOING RETAINER · TARGETED BRIEFINGS

GLSD.AI

daniella@globallenspodcast.com

www.ingramcontent.com/pod-product-compliance
Lightning Source LLC
LaVergne TN
LVHW011053110826
845149LV00015B/3482

* 9 7 9 8 9 9 5 9 0 1 3 1 0 *